1 MONTH OF
FREE
READING

at

www.ForgottenBooks.com

By purchasing this book you are eligible for one month membership to ForgottenBooks.com, giving you unlimited access to our entire collection of over 1,000,000 titles via our web site and mobile apps.

To claim your free month visit: www.forgottenbooks.com/free913783

ISBN 978-0-265-94601-5
PIBN 10913783

This book is a reproduction of an important historical work. Forgotten Books uses
state-of-the-art technology to digitally reconstruct the work, preserving the original format
whilst repairing imperfections present in the aged copy. In rare cases, an imperfection in
the original, such as a blemish or missing page, may be replicated in our edition. We do,
however, repair the vast majority of imperfections successfully; any imperfections that
remain are intentionally left to preserve the state of such historical works.

.

The University of Chicago
FOUNDED BY JOHN D. ROCKEFELLER

THE DEVELOPMENT OF THE CENTRAL CYLINDER OF ARACEAE AND LILIACEAE

A DISSERTATION

SUBMITTED TO THE FACULTY OF THE OGDEN GRADUATE SCHOOL OF
SCIENCE, IN CANDIDACY FOR THE DEGREE OF
DOCTOR OF PHILOSOPHY

(DEPARTMENT OF BOTANY)

BY

MINTIN ASBURY CHRYSLER

CHICAGO
1904

BOTANICAL GAZETTE

SEPTEMBER, 1904

THE DEVELOPMENT OF THE CENTRAL CYLINDER OF ARACEAE AND LILIACEAE.

CONTRIBUTIONS FROM THE HULL BOTANICAL LABORATORY. LXII.

MINTIN ASBURY CHRYSLER.

(WITH PLATES XII–XV)

THE fundamental unity of the vascular structures found in the higher plants was perceived by VANTIEGHEM, whose conception of the stele and its modifications, outlined in 1886 (**17**), displaced the earlier view of DEBARY. But it became apparent subsequently that VANTIEGHEM'S assumptions were not sufficiently supported by observation. For example, it was shown by GWYNNE-VAUGHAN (**3**) that polystely does not arise by bifurcation of the protostele in the genus Primula, and JEFFREY (**6**) proved the same for *Pteris aquilina*. VANTIEGHEM'S theory is also open to the objection that it is founded on the conditions occurring in a highly organized group of plants, while there would seem to be a better prospect of finding a primitive condition of the vascular system among pteridophytes. In 1897 JEFFREY (**5**) proposed a stelar theory in which this objection is met, the essential feature of which is the important influence on the central cylinder of the outgoing leaf or branch traces. Emphasis is also placed on the study of the young vascular axis, on account of its recognized importance in accordance with the principle of recapitulation. The following diagrams may serve to show the main differences between the two theories:

I. VanTieghem.

1. Protostele

2. Medullated monostele   4. Polystele

3. Astele (schizostele)   5. Gamostele

II. Jeffrey (using the same numbers to designate equivalent types).

(1) Protostele

(5 and 4) Siphonostele with internal phloem and endodermis (amphiphloic siphonostele)

(3) Siphonostele with internal endodermis (ectophloic siphonostele)

(2) Siphonostele without internal phloem or endodermis.

It will be noticed that Jeffrey derives the vascular structures characteristic of the seed plants from those of the pteridophytes by a process of reduction; further, he considers the pith to be simply fundamental tissue which has intruded through the foliar or ramular gaps, while VanTieghem assumes a stelar origin for the pith.

The researches of Jeffrey (7) and Gwynne-Vaughan (4) seem to place beyond question the view that the ferns possess an amphiphloic siphonostele derived from a protostelic condition by the bending in of phloem, endodermis, and cortex above the point of exit[1] of the foliar traces; but that the seed plants have primitively a central cylinder built on this plan is a generalization which must be tested by the examination of representatives from a number of typical families in different regions of this great group. With this object in view I have undertaken, at the suggestion of Dr. Jeffrey, to investigate the development of the central cylinder in two characteristic monocotyledonous families, the Araceae and the Liliaceae. Such an investigation ought to answer the following questions:

1. What bearing on current stelar theories has the development of the central cylinder in these families?

2. Are the amphivasal bundles found in so many monocotyledons to be considered a primitive type?

[1]For the sake of clearness the leaf traces will be treated as if they originated at the central cylinder, regardless of the actual direction of their development, which in most cases has not been made out.

3. Does the structure of the young stele throw any light on the question of the origin of the monocotyledons?

ARACEAE.

The number of forms which have been available in this inquiry has not been large, but they are sufficiently varied in their affinities, and appear to the writer to yield no uncertain result.

Pothoideae. This subfamily is regarded by ENGLER (1) as the most primitive one in the family. Hence one of its most available representatives will be first described.

ACORUS CALAMUS.[2]—In a seedling of this plant the central cylinder in its lowest region is a solid mass of vascular tissue, consisting of a core of xylem and a ring of phloem, surrounded by pericycle and endodermis, that is, it is a typical protostele. One trace is given off to the cotyledon, and usually three traces to each of the next three or four leaves, after which the number of foliar traces is increased. In the region where the traces of the second leaf are given off, the central cylinder is seen to possess a parenchymatous pith, which is continuous with the pericycle through the gaps in the vascular tissues caused by the bending out of the traces (*fig. 1*, which, however, represents a higher region of the stem). The endodermis does not bend inward through the gap with the pericycle, but remains unbroken, a portion of it surrounding the trace as it passes outward. Followed downward through the stele the pith either becomes narrower and disappears above the point of exit of the cotyledonary trace, or in some cases enlarges at this point and communicates with the pericycle at the higher node; followed upward the pith widens out with the enlarging central cylinder. As the three traces of the third leaf bend outward, the pith again communicates with the pericycle; since the median trace is the largest of the three, the gap it leaves in the vascular ring is the widest; in fact, the gap of one or both of the lateral traces may be filled only by a single row of parenchymatous cells or may not be present. Up to this point the vascular ring is practically continuous, owing to the foliar gaps being so short, but in the higher regions of the young stem the gaps remain open longer, so that the central cylinder appears to be made up of a ring of sepa-

---

[2] The nomenclature employed in this paper is that of Engler and Prantl.

rated bundles which are at first collateral, but soon become amphivasal. *Fig. 1* shows this region of the stele. Not until a considerably older stage is reached do certain bundles turn inward and run for a distance in the medulla before turning out to the leaves. *Fig. 2* shows part of a section through the mature rhizome; most of the bundles are amphivasal, and some of them run in the medulla; *g* is a gap through which a medullary bundle has lately passed, and it will be noticed that the endodermis curves inward around the edges of the gap for a short distance, thus making the cortical parenchyma continuous with that of the medulla. This intercommunication of cortex with medulla is even better marked in the base of the flowering axis, as is shown in *fig. 3*. It will be seen that the endodermis extends around the edges of the gaps for a short distance, and completely encircles one small section of the vascular ring. It seems reasonable to believe that if the gaps in the central cylinder of the seedling of Acorus were not so narrow the cortex might communicate with the pith as it does in the seedlings of other Araceae possessing a wider central cylinder.

ANTHURIUM ACAULE.—In the hypocotyl the central cylinder is a hollow tube consisting of xylem, phloem, pith, and surrounded by an endodermis. Just below exit of the single cotyledonary trace the vascular ring breaks up into a circular row of five or six collateral bundles or meristeles. Above the exit of the cotyledonary trace the row is horseshoe-shaped, but soon becomes circular again owing to the reunion of the bundles on the two sides of the cotyledonary gap. The endodermis cannot be followed clearly owing to its poor development. The stele retains its form of a ring of about six collateral bundles through the first internode. At the upper end of the internode several bundles divide, and certain of these turn outward as traces of the second leaf, while others turn inward and run upward through the pith, becoming traces of leaves higher up. In the young stem no concentric strands have been found.

Monsteroideae.—MONSTERA DELICIOSA.—The hypocotyledonary stele consists of a circle of collateral bundles inclosing a parenchymatous medulla. Nearly a third of these bundles bend outward at one side to supply the cotyledon; a little above this point bundles begin to run in the medulla; in other words, the central cylinder early assumes the characters seen in the adult stem. The bundles are collateral in all parts of the stem. No endodermis can be distinguished.

Calloideae.—SYMPLOCARPUS FOETIDUS has already received some attention in JEFFREY'S preliminary studies of monocotyledons (6). The seedling at the age of one year consists of a spherical tuber about 1^{cm} in diameter; from the upper side of this rises a conical bud with a cylindrical base 4^{mm} in diameter, from which spring several roots. A transverse section through the basal region of the tuber shows an elliptical row of collateral strands, each surrounded by an endodermis (*fig. 4*). A little higher up several bundles at one side of the ellipse turn outward, so that at about the middle region of the tuber the bundles are arranged as a horseshoe. Opposite the open part of the horseshoe there is frequently a swelling of the tuber, and in some cases this part of the tuber separates off at a slightly higher level by an absciss layer; this part accordingly constitutes the cotyledon, and the opening in the central cylinder is the cotyledonary gap. Toward the upper part of the tuber the separate strands approach one another, as is shown in *fig. 5*.[3] At *g* is the cotyledonary gap; most of the vascular strands have fused laterally, producing a hollow vascular cylinder with an external and internal phloeoterma (using the term in STRASBURGER'S sense (15, p. 310)), broken by the wide cotyledonary gap and by several areas where the individual bundles have not yet fused; through these openings the external and internal phloeotermas are obviously continuous. The latter may persist for some distance upward, finally becoming indistinguishable, or may degenerate quite early, and, as is seen in *fig. 7*, the external phloeoterma runs for a short distance around the edges of the cotyledonary gap, and then disappears. Compare *figs. 2* and *3* of Acorus, also JEFFREY'S figure of *Ranunculus rhomboideus* (6, *fig. 16*). A little higher up the cotyledonary gap closes and the stele forms a hollow tube with external and internal phloeoterma. Almost immediately, however, the vascular tissue aggregates into separate strands, the xylem of which is disposed circularly (amphivasal bundles), and a few of these turn into the central region of the stele (*fig. 6*). Each of these bundles is surrounded by a portion of the internal phloeoterma, if this has not

[3] *Figs. 5, 6, 7, 8*, and *11* are from sections treated with sulfuric acid.

already become invisible in this region of the stem, as is seen to be the case in the stem represented in *fig. 8.* These bundles soon become quite numerous and run upward for some distance before resuming their collateral structure and passing outward to the leaves.

Two points of interest in this plant are the existence of a well-marked internal as well as external phloeoterma in the young stem, and the early disappearance of the internal phloeotermal layer. It can hardly be doubted that the thin-walled tissue forming the pith of the stele is simply extrastelar tissue which enters at the base of the stelar system, and through the cotyledonary gap. Absence of the protostelic condition is probably to be accounted for by the shape of the stem; it is in the region of the cotyledonary gap that the central cylinder shows its most primitive condition, namely a vascular tube possessing both external and internal phloeoterma. The spaces between the vascular segments in the basal region of the stem are not foliar gaps, as JEFFREY's account seems to imply (**6**, p. 29), for this region of the stem is the hypocotyl, and further there are no outgoing bundles between the segments referred to. Separation of the segments may be due to expansion of the young stem as it assumes its tuberous shape.

CALLA PALUSTRIS has been sufficiently described by JEFFREY (**6**). The development of its stele follows pretty closely the course outlined for Acorus, though the endodermis does not seem to be well developed in Calla, also the foliar gaps extend for a greater distance than is the case in Acorus.

Philodendroideae.—SCHIZMATOGLOTTIS ROEBELINII seems to show a scattered disposition of the vascular strands in all parts of its seedling. The material available has not permitted a satisfactory study.

PELTANDRA VIRGINICA.—The seedling possesses a tuberous base consisting of a somewhat cylindrical axial portion with a thick cotyledon applied to its side; the cotyledon is separated in its upper part from the axial portion by a prominent absciss layer. In *fig. 9, c* represents the cotyledon, and *r, r* secondary roots. A section through the basal part of the tuber shows about eight collateral bundles arranged in a circle (*fig. 10*). Most of these bundles are given off to the cotyledon, so that only a few slender strands continue the upward course; toward the upper part of the tuber these enlarge and each is seen to be provided with an endodermis whose cells show a cutinized band girdling the radial walls. The strands now unite laterally into a flattened arch whose hollow is turned toward the cotyledon; by continued increase in the vascular tissue the arch becomes more and more nearly a complete circle. In *fig. 11* the cotyledon lies to the right; the individual sheaths have fused to form a common endodermis which is continuous outside and inside the arch; *r* is the trace of a root, which as usual leaves no gap in the central cylinder. *Fig. 12* shows the central cylinder at a slightly higher level; the opening to the right faces the cotyledon, and is undoubtedly the cotyledonary gap. Soon this closes entirely, and at this level the vascular tissue of the stele becomes partly broken up into separate strands, some of which turn into the medulla; each strand and segment of the stele possesses its own endodermis. Amphivasal bundles are found at this level and in the later formed regions of the stem, but they are not so characteristic of Peltandra as of Symplocarpus, to which plant Peltandra evidently possesses many resemblances with respect to its central cylinder. The medullary strands are connected with the traces of all leaves above the cotyledon, and each trace leaves the central cylinder through a gap, around the edges of which the external and internal endodermis are continuous. Eventually, however, the endodermis becomes obsolete, and an increase in the number of medullary strands gives the stele the appearance characteristic of monocotyledons generally. It should be mentioned that the ring of bundles is not always present in the lower part of the tuber; in such cases bundles are so poorly developed in this region that a central cylinder cannot be said to exist below the cotyledonary gap.

ZANTEDESCHIA AETHIOPICA (the ordinary calla lily) and Z. ALBO-MACULATA may be described together, since the seedlings are very similar. As the stele of the root merges into that of the hypocotyl it assumes a pith into which several strands turn from the original vascular ring, and soon the whole stele is converted into a network of anastomosing strands. From this network about six bundles are given off to the cotyledon, whose base forms a sheath around the younger leaves. In the succeeding regions of the stem the bundles

pursue the course ordinarily seen in a monocotyledonous stem.   In
no part of the stem is an endodermis well developed.

Colocasioideae.—ALOCASIA ODORATA.—Above the point of exit
of the cotyledonary traces the stele is represented only by a scanty
vascular mass of flattened form, its side being turned toward the
cotyledon.   Further upward this mass splits into several strands,
and a ring-shaped row of bundles is completed by the appearance *de
novo* of several delicate strands between those already present and
the cotyledon.   The flattened vascular mass referred to seems to
represent the same condition as that shown for Peltandra in *fig. 11*,
namely, there is an unusually wide cotyledonary gap, which is not
closed in the ordinary way owing to the tendency throughout the plant
for the vascular strands to lie widely separated.   In a slightly higher
region of the stem several bundles come to lie in the medulla and some
of the bundles assume the amphivasal shape.   No endodermis was
found in any part of the stem.

CALADIUM BULBOSUM.—Departure of the cotyledonary trace
causes no break in the narrow stele of the seedling; the stele soon
becomes complicated by medullary strands which anastomose with
one another.   In many sections, however, it may be seen that the
cortex communicates freely with the medulla above the point of exit
of a leaf trace.   No endodermis has been demonstrated.

Aroideae.—ARUM ITALICUM.—The five traces which pass into the
sheathing base of the cotyledon arise from a complex vascular mass,
and the succeeding traces run for a short distance in the medulla.
The peculiar habits of sprouting described for another member of the
genus by RIMBACH (**10**) and SCOTT and SARGANT (**13**) have probably
had the effect of modifying the vascular system; and no part of the
plant suggests a primitive condition, but on the contrary a highly
specialized one.

ARISAEMA TRIPHYLLUM.—The method of sprouting is essentially
like that of Arum.   The five cotyledonary traces rise from a vascular
mass whose elements anastomose in a complex manner.   Above this
region the bundles pursue a more nearly vertical course, but are not
arranged in a definite central cylinder surrounded by endodermis.
In older seedlings the bundles form an extensive network in the
central region of the corm.   It is probable that the phylogenetic

development of this corm has been accompanied by considerable changes in the vascular system, leading to complications which render this plant unsuitable for the purposes of the present inquiry.

ARISAEMA DRACONTIUM, A. SPECIOSUM, A. INTERMEDIUM, and A. TARTARINOWII all resemble *A. triphyllum* in having seedlings which show a complex network of bundles. They all likewise produce a corm.

TYPHONIUM DIVARICATUM has a vas͗   ͏tem so similar to that found in Arisaema that it does not meri͟   rate description.

In viewing in a general way the genera   ͏ described the question arises: What characters are to be reg͟   ͏s primitive? The
answer must be, those which occur in the fi͟.   ͏part of the stem,
unless there is reason to believe that this regi͟   ͏een influenced
by the assumption of some special habit, such a͟   tuberous or
bulbous habit. The stem of Acorus is relatively free͟   ͏ external
influences, on account of its geophilous habit; its central ͟͏inder is
at first protostelic, then siphonostelic with a pith communicating with the pericycle through the foliar gaps. Judging from the conditions in Symplocarpus and in the mature organs of Acorus, we may infer that if the central cylinder of the Acorus seedling were not so narrow the endodermis and cortex might here also enter through the gaps, in which case the stele would differ from that characteristic of the ferns mainly in the absence of internal phloem, a feature which appears to be quite rare in seed plants. The simple siphonostelic stage persists in Acorus for several internodes, and the stem looks much like that of a dicotyledon; higher up some segments of the stele become amphivasal, and this may be regarded as the first appearance of a monocotyledonous character; very soon certain strands begin to run in the medulla, and so the monocotyledonous nature of the stele is established. The steles of the various genera differ from the type just described in a modification of the basal part of the stele in accordance with the tuberous habit, as in Symplocarpus, or in the rapid disappearance of the phloeoterma, as in Peltandra, or in the early appearance of the medullary strands, as in Arisaema. Whatever may be the nature of the pith in Acorus, there seems to be good reason for believing that in Peltandra and Symplocarpus the pith is

simply fundamental tissue which has been inclosed by the gradual curving around of the edges of the cotyledonary gap until they meet; moreover, the stele is in open communication with the cortex in the basal region of the tuber. It has been shown that in Symplocarpus the internal phloeoterma undergoes a more or less early degeneration, so that the included parenchyma comes to lie next to the xylem, and so might be mistaken for stelar tissue in the upper part of the stem.

It is of interest to note that from a general morphological study of Araceae ENGLER (1) places Acorus among the most primitive genera of the family, and members of the Aroideae, such as Arum and Typhonium, among the most highly developed of the family. My observations on the seedlings accord in the main with ENGLER'S classification; the young stem of Acorus possesses a simple stele, while members of the Aroideae early acquire the most complicated vascular system found in the family.

LILIACEAE.

In this family over fifty species have been studied, representing all the large subfamilies; in most cases both the adult plants and seedlings have been examined. The search for primitive types has convinced me that ancestral characters are most likely to be preserved in a rhizome, since such a stem is free from the modifying influences of an aerial life; hence the first subfamily to be treated is one in which most of the members have the basal portion of the stem a rhizome.

Asparagoideae.—CLINTONIA BOREALIS.—The plant is characterized by a horizontal rhizome which turns upward at the end, bears a number of scales and several foliage leaves, and terminates in a scape carrying an umbel of flowers. *Fig. 13* represents a cross section through the rhizome a short distance before it turns upward; it will be seen that the stele forms a tube perforated at the point of exit of several leaf traces, also that there are no medullary bundles. *Fig. 14* is a more highly magnified view of a portion of the stele. It shows that an internal as well as an external phloeoterma is present, and that these are continuous through the foliar gap; also that some of the meristeles are amphivasal.

CLINTONIA UMBELLATA.—A cross section through the rhizome is shown in *fig. 15*; two foliar gaps with their traces are to be seen, and here also is an internal as well as an external phloeoterma. Certain of the strands are amphivasal, and three of these have left the stelar ring and run immediately adjacent to it in the medulla. The vascular system of this plant evidently represents a condition slightly more complicated than that present in *C. borealis*. Unfortunately seed- . lings of neither species of Clintonia have been available, so that the origin of the interesting condition seen in the mature stem remains unknown.

MAIANTHEMUM BIFOLIUM.—The general habit of the plant is similar to that described for Clintonia, though the two foliage leaves arise from a higher region of the aerial shoot. *Fig. 16* represents a section through the rhizome a short distance above where it turns upward. The heavily cutinized external phloeoterma is a prominent feature, and inside of it is a circle of collateral bundles; three amphivasal bundles run in the medulla, but these do not become leaf traces; on the contrary they end as they begin, namely, by joining bundles of the vascular ring. Throughout the horizontal course of the rhizome no medullary bundles are present. Several leaf traces are to be seen at various stages in their escape from the stele; it will be noticed that they cause no break in the continuity of the phloeoterma.

In the seedling the stele contains pith even in the hypocotyl; as the single cotyledonary trace leaves the stele, the phloeoterma bends inward around the edges of the gap, but does not lose its continuity; the pericycle is continuous with the pith through the gap and no amphivasal strands are present. To the second leaf three traces are given off; the median trace causes the phloeoterma to bend inward, as does the cotyledonary trace (*fig. 17*); the lateral traces emerge exactly as in the adult stem (*fig. 16*). The third leaf receives three traces which leave the stele as do those of the second leaf; the same is true of the fourth and fifth leaves. Comparing the stele of this plant with that of *Clintonia borealis*, the absence of internal phloeoterma and the presence of amphivasal medullary strands in the former are to be noted, though these do not make their appearance until a late stage of development.

SMILACINA STELLATA.—As the primary root merges into the hypocotyl, the stele becomes hollow and the vascular tissue aggregates in several collateral strands at the periphery of the stele. An external

phloeoterma is present and is not broken by the exit of a large strand
of vascular tissue to the cotyledon. *Fig. 18* shows the appearance of
the stele in the first internode, and illustrates the tendency which the
stele has to break up into separate strands. The three traces of the
second leaf arise in the same manner as the cotyledonary trace; above
this level, however, some of the strands become concentric, and one
or two branches are given off into the pith, where they run only a
short distance, join bundles of the vascular ring, and then pass out
to leaves. *Fig. 19* shows such a strand at *m*, and also a leaf trace (*t*)
which is just leaving the vascular ring. Higher up other medullary
strands run for a greater distance in the pith and turn outward to
leaves without anastomosing with bundles of the vascular ring. The
seedling of this plant shows clearly the gradual appearance of monocotyledonous characters in a central cylinder which in its first formed
part closely resembles that of a dicotyledon. The mature stem both
in its subterranean and aerial regions differs from the rhizome of
Maianthemum in having a number of medullary strands.

SMILACINA RACEMOSA.—In the youngest plants obtainable the
central cylinder exhibited the characters of the adult stem, that is the
bundles are scattered through the medulla. Seeds of this plant
failed to sprout.

STREPTOPUS ROSEUS.—The stele of the seedling resembles that of
Smilacina stellata in that it consists of a ring of collateral bundles
surrounded by a phloeoterma and enclosing pith; but the bundles
early become concentric and afterwards some of them run in the
medulla. Bundles of the vascular ring turn outward as leaf traces
without destroying the continuity of the phloeoterma.

POLYGONATUM BIFLORUM and P. VERTICILLATUM.—The central
cylinder, at first a solid mass, becomes divided into about six widely
separated collateral strands at a comparatively young stage; later
several medullary bundles appear; a phloeoterma is not distinguish-
able. The wide separation of the strands is probably due to the fact
that the subterranean stem (a horizontal rhizome) early becomes
swollen into ovoid form through deposition of starch.

MEDEOLA VIRGINIANA.—The cotyledonary gap is wide; as this
closes the stele becomes divided into six or eight meristeles arranged
circularly, each provided with its own endodermis and having a

collateral structure. The meristeles unite at the next node and a single strand turns outward, leaving a wide gap in the vascular ring. Again the meristeles separate widely, owing no doubt to the fleshy nature of the stem which by this time has begun to show its habit of a horizontal somewhat swollen rhizome. As the stem turns upward into the air the meristeles approach one another and some of them become amphivasal. At about this point the internal endodermis disappears, but the external layer becomes strongly cutinized. The amphivasal strands resume their collateral structure at a slightly higher level; medullary strands are absent in most plants.

TRILLIUM GRANDIFLORUM.—The subterranean stem is a vertical rhizome which becomes thicker and more ovoidal as the plant grows older, owing to deposition of stores of starch. In the young stem the central cylinder is a solid mass of vascular tissue for a few internodes. The first leaves have three traces; the median trace is much the largest, and as it leaves the central cylinder the latter becomes somewhat crescent-shaped; the lateral traces are very delicate and by their departure leave no indentations in the stele. After exit of the traces of the third or fourth leaf, however, there is intrusion of fundamental tissue into the central cylinder, since the angles of the crescent above referred to curve around and finally close in on the side next the trace. This condition is shown in *fig. 20*; t is the median trace, t_r is one of the lateral traces whose gap was narrower than that of the median trace and had already closed at this level. The writer fails to see how the thin-walled tissue inside the stele can be regarded as anything but a portion of the fundamental tissue inclosed by approximation of the vascular tissue at the sides of the gap. The appearance above described may be masked by the overlapping of two foliar gaps on opposite sides of the stele; in such a case the stele is broken into two halves (*fig. 21*), a condition which has frequently been described for various ferns. In the upper part of the rhizome the stele becomes complicated by bridges of vascular tissue reaching from one side of the stele to the other; also certain of the leaf traces run for a short distance in the medulla before turning outward. It should be remarked that these medullary strands are amphivasal. In some seedlings the stele retains its solid or protostelic character for many internodes, and it is possible that the diversity noticed in the various serial sections of seedlings may be due to some of them belonging to *T. erectum* rather than to *T. grandiflorum*, since the seedlings of the two species are hard to distinguish in the field, and I have been unsuccessful in attempts to grow them from seed. In many instances a leaf trace arises from one side rather than from the base of the gap, as has been observed in many ferns, or the trace may run vertically for some distance before turning out from the stelar ring. This condition is shown by the trace marked t_1 in *fig. 22*; t_2 and t_3 are the lateral traces of the same leaf; the last has not quite broken away from the stele.

TRILLIUM SESSILE and T. RECURVATUM greatly resemble *T. grandiflorum* in the seedling stage, but differ from the last species in having wider gaps and showing concentric bundles in a younger part of the stem.

ASPARAGUS OFFICINALIS, A. VERTICILLATUS, A. SPRENGERI, A. VERTICILLATUS, A. BROUSSONETII, and A. MEDEOLOIDES do not appear to throw any light on the problems under consideration on account of the complications attending the formation of lateral buds.

RUSCUS ACULEATUS has a seedling much resembling those found in the genus Asparagus. In the seedlings available the stele had already assumed its mature condition.

Dracaenoideae.—YUCCA FILAMENTOSA.—The hypocotyledonary stele consists of a hollow vascular tube from which about one-third of the vascular tissue turns outward to the cotyledon, leaving a U-shaped stele whose pith is in free communication with the cortex. No phloeoterma was observed. Almost immediately strands turn from the U into the pith, and before the cotyledonary gap is closed these medullary strands are quite numerous. These become traces of higher leaves, so that the stele in this plant quickly attains the characteristic monocotyledonous condition. Nearly all the vascular strands are collateral.

YUCCA ANGUSTIFOLIA and Y. BACCATA resemble *Y. filamentosa* in the young state; in the first-named species the medullary bundles are somewhat later in arising than in the two other species.

DRACAENA DRACO, D. RUBRA, D. VEITCHII, and CORDYLINE AUSTRALIS differ in no essential respect from Yucca as regards the development of the stele.

ASTELIA sp. (*Funkia coerulea*) offers no points of significance.

Lilioideae.—LILIUM CANADENSE.—The young plant consists of a vertical axis upon which is set a spiral series of fleshy awl-shaped scales which are loaded with starch; to each of these three traces run from the tubular central cylinder, taking a course directly outward or even curving downward for a short distance after leaving the central cylinder. These traces, though slender, subtend foliar gaps which frequently extend the whole length of an internode, so that the central cylinder has the appearance of three separate collateral strands, except at the nodes, where a vascular ring is formed, and in the lower part of the seedling, where the scales are more crowded. *Fig. 23* shows the appearance of the central cylinder at a node, t is the median trace; bordering the vascular strands are cells differing from the surrounding parenchyma by their entire lack of starch; these may represent a phloeoterma. In the higher regions of the stem the usual medullary bundles appear, and some of these are amphivasal.

ERYTHRONIUM AMERICANUM, CALOCHORTUS VENUSTUS, GALTONIA CANDICANS, SCILLA HYACINTHOIDES, CAMASSIA FRASERI, HYACINTHUS CANDICANS, and LACHENALIA PENDULA early assume the bulbous habit characteristic of the adult plant, hence the stem is flattened in the vertical direction. The complications produced in the vascular tissues by this habit render these genera unprofitable for study, and since there is no reason to believe that the bulbous condition is a primitive one, no description of these genera will be necessary here.

Allioideae.—ALLIUM CEPA, A. CANADENSIS, and A. ANGULOSUM have seedlings much resembling those of the last group in their bulbous habit and intricate vascular system.

AGAPANTHUS UMBELLATUS has a stele much resembling that of Allium.

Asphodeloideae.—ASPHODELUS FISTULOSUS, ASPHODELINE LIBURNICA, BULBINE ANNUA, B. FRUTESCENS, ANTHERICUM LILIAGO, CHLOROPHYTUM ELATUM, KNIPHOFIA TYSONI, K. BREVIFOLIA, and ALOE sp. agree in having short internodes and passing quickly through the early stages of stelar development, so that the medullary bundles are found near the exit of the cotyledonary traces. Further, an endodermis is rarely discernible, so that these genera are unsuitable in the present investigation.

ANEMARRHENA ASPHODELOIDES has been studied with much interest because Miss SARGANT (11, 12) considers that the vascular system of the seedling represents a primitive type. The theory of this author considers chiefly the cotyledonary traces and their insertion; it is natural to inquire whether the stele of the older seedling shows features which may be regarded as primitive. The stele possesses a medulla below the exit of the two cotyledonary traces; these subtend wide gaps through which the cortical and medullary parenchyma freely communicate; the traces of the second leaf are three in number, and before they emerge medullary bundles have made their appearance. Except in the root an endodermis cannot be identified. This fact, and the early appearance of medullary strands, and the presence of a pith in the hypocotyl I do not regard as primitive characters, though it is evident that a plant may retain some ancestral features and lose others, so that the disposition of the cotyledonary strands may still represent an ancestral type.

Melanthoideae.—GLORIOSA SUPERBA.—The peculiar habit of the subterranean portion of the stem has been fully described by QUEVA (9), and sufficiently accounts for the complications found in the lower part of its stele; in the upper internodes of the seedling, however, the vascular strands are arranged in a simple ring, and certain of the strands turn outward as leaf traces after anastomosing with adjacent members of the ring.

UVULARIA GRANDIFLORA.—At the point of departure of the cotyledonary trace a wide gap is left in the vascular tissues; here fundamental tissue enters and extends downward into the hypocotyl for a short distance as well as upward. *Fig. 24* shows the stele at level of the cotyledonary gap; the cotyledonary trace is not visible because it bends downward after leaving the stele; though no distinct phloeoterma is visible, the small-celled tissue surrounding the stele certainly does not seem to be continuous across the gap, as it is in some of the adult stems already described. At one place in the stele it will be noticed that the xylem surrounds a mass of phloem, so that the concentric bundles begin to show themselves at this early stage; they make up the whole vascular ring above the point of exit of the second leaf trace; some of them then turn inward and run in the medulla, but here they soon become collateral.

Viewing in a comparative way the genera of Liliaceae described in the foregoing paragraphs, it appears that Trillium exhibits very clearly the stages in development of the stele. These stages may be briefly enumerated as follows: (1) the protostelic condition is present in the basal part of the stem and persists through one or more internodes; then follows (2) the siphonostelic condition in which cortical tissue is included in the stele above the point of exit of the leaf traces and thenceforth forms a medulla; (3) many segments of the stele take on the amphivasal character; (4) strands of vascular tissue, usually amphivasal, turn into the medulla where they run for a greater or less distance and may become connected with leaf traces. Though the stem of Trillium seldom shows any traces of a phloeoterma, *Clintonia borealis* presents a diagrammatic example of a stele which never gets beyond stage (3), and has external and internal phloeoterma which communicate through the foliar gaps. The internal phloeoterma is probably degenerate in Maianthenum except at the edges of the leaf gaps of the young stele; there may be a physiological correlation between the very heavily cutinized external layer and the absence of an internal layer; stage (4) is much delayed in this plant. In *Smilacina stellata* stages (3) and (4) appear sooner; the phloeoterma is less distinct. Medeola and Lilium show the effect of long internodes combined with extended gaps in breaking up the central cylinder into several strands arranged on the circumference of a circle. Uvularia and Streptopus quickly pass into stage (3). Many members of the family such as Allium have assumed the bulbous habit, and in the very short stem of these plants the medullary strands appear very early. They probably express the highest order of specialization shown in the family.

As to the bearing of the foregoing observations on the central cylinder of the two families upon so-called stelar theories, it may at once be stated that though the pteridophytes must be the critical group in any discussion of these theories, yet information from even so highly specialized a group as the monocotyledons is of importance, if we acknowledge the descent of the seed-plants from fern-like ancestors. Many of the Liliaceae studied do not seem to afford any evidence on the points in dispute, which is not to be wondered at when one considers the adaptations which these plants show; but several plants of both families show characters which, to say the least, are significant.

Concerning VanTieghem's types, the polystele need not be considered, for no monocotyledon has yet been found with internal ·phloem; the medullated monostele may be present in such forms as Acorus and Smilacina, but the condi ion may be equally well explained by assuming the degeneration of an internal phloeoterma, deriving this condition from that shown in Clintonia; what may be called an astele or schizostele is probably present in the mature stem of most members of the two families, but in none of the cases examined does t arise by the breaking up of the stele followed ·by the uniting of the broken ends of the external endodermis on the inner side of the meristeles; on the contrary, wherever the endodermis is discernible in the region of splitting up of the stele, there is an internal as well as external endodermis which communicate at the leaf gaps (*e. g.,* Clintonia); in Symplocarpus each strand which turns into the medulla is surrounded by a portion of the internal endodermis.

Turning to the theory of Jeffrey, a consideration of the figures which accompany this paper shows that there is evidence in the case of the two families studied to support his fundamental statement that the siphonostelic type of central cylinder "is primitively a fibro-vascular tube with foliar lacunae opposite the points of exit of the leaf traces" (6, p 38). That the simple tubular condition is found for only a few internodes in most cases is due to the monocotyledons having acquired a new mode of insertion of the leaf traces, which has replaced the mode characteristic of ferns. However, in rhizomes, whose subterranean position has shielded them from the disturbing effects of aerial life, a more primitive type of stele is frequently found; seedlings almost universally show a gap in the central cylinder above the point of exit of the cotyledonary trace, un'ess indeed they are protostelic at this level, as in Trillium. The siphonostelic nature of the central cylinder is often retained for several internodes, but sooner or later the medullary strands appear, or the gaps persist through an entire internode, in either case resulting in the masking of the essentially tubular nature of the stele. That the young central cylinder of so highly organized a group as the monocotyledons shou'd

show fern-like characters is a fact of considerable phylogenetic significance. The evidence concerning the extrastelar or intrastelar
origin of the pith is not so plain, but from the method of closure of
the cotyledonary gap in Peltandra and Symplocarpus and of the foliar
gaps of Trillium, I am led to believe that the tissue in question has
been included; the hypocotyledonary region in Peltandra and Symplocarpus also suggests the unity of extrastelar and intrastelar tissues.
Narrowness of the gaps would account for the failure of the endodermis and cortical tissue to enter through the foliar gaps in Acorus,
and the absence of internal endodermis in such plants as Maianthemum may be ascribed to degeneration of such a layer as is found in
the lower part of the stele in Symplocarpus but disappears in the
higher regions of its stele. Thus it appears that the terms "cortex"
and "pith" should be used only in a topographical sense, and not as
implying a difference of origin, for morphologically they must be
regarded as identical, as regions of the "fundamental tissue," using
this term in the sense of SACHS and DeBARY. Hence if the term
"stele" is used, it should be restricted to the vascular elements of the
central cylinder, as is insisted on by FARMER and HILL (2). Further,
the researches of SCHOUTE (14) have shown that HANSTEIN's derma-
togen, periblem, and plerome do not correspond to VanTIEGHEM's
epidermis, cortex, and stele, so that there no longer appears to be
any necessity for postulating a common origin for all the tissues found
inside the endodermal ring. On the whole, then, the development of
the stele in the two families in question appears to support the gener-
alizations made by JEFFREY.

MEDULLARY BUNDLES.—The writer is inclined to believe that
these did not originate as leaf traces, but as strands to which leaf
traces subsequently became attached. This tentative view rests upon
the following considerations:

1. The tendency of the monocotyledonous stele to break up into
segments makes it easy for a strand to leave its vertical course at the
periphery of the stele and run for a distance in the medulla; such a
strand may at a higher level return to its original course, or may join
the stelar ring at the opposite side. Both of these conditions are to
be seen in the young stele of *Smilacina stellata*. In Maianthemum
the first medullary strands to appear do not come in contact with the

leaf traces which arise in this region (*fig. 16*), but higher up return to the stelar ring. It is probable that the anastomosing strands seen in Trillium, Zantedeschia, etc., are of the same nature.

2. VanTieghem (**16**, p. 172) has shown that in *Acorus gramineus* after a bundle has run for a distance in the medulla it divides, one part bending outward as a leaf trace, the other pursuing the medullary course, again dividing further on, and finally passing out to a leaf. The bundle marked *b* in *fig. 2* is in process of division into a medullary bundle and a leaf trace.

3. Medullary bundles are either absent or few in number in rhizomes, but become numerous as soon as the stem turns upward into the air; this is not altogether due to the greater development of leaves in the aerial part.

It is probable that these strands have an important mechanical function, which may explain their paucity in rhizomes; they can hardly have arisen in consequence of a crowding out into the medulla of the too numerous vascular elements of the stelar ring, for they occur in stems whose meristeles do not form a complete ring, *e· g.*, *Smilacina stellata* (*fig. 19*).

Amphivasal bundles. The mode of formation of these was observed by VanTieghem in the mature stem of *Acorus gramineus* (**16**, p. 171); I have traced their formation in the young stele of *Acorus calamus* and *Smilacina stellata*. Starting with a simple collateral bundle of the vascular ring it may be seen that the tracheids increase in number so as to give the xylem a U form and finally an O form. Some strands never go any further than the U stage, and some that have become concentric lose the tracheids of their outer side. It is plain, then, that amphivasal bundles are derived from collateral ones and are simply a modification of the latter type. Since phylogenetic significance has been attached to the concentric and mesarch bundles found in the petioles and peduncles of cycads, it has been thought worth while to find in what parts of the plant in the Araceae and Liliaceae the amphivasal bundles occur. The result of a somewhat extensive investigation of this point may be briefly stated as follows: (1) only collateral strands are found in the lowest part of the stem of the seedling; (2) amphivasal strands are found in the older stem in nearly every genus; (3) the floral axes show only collateral strands, which may be arranged in a circle or scattered; (4) only collateral strands are found in the leaves. Hence amphivasal strands are to be regarded as cenogenetic structures.

The observations recorded in this paper seem strongly to support the statement made by JEFFREY (8) that neither the medullary course of the bundles nor their amphivasal nature are primitive features, but that they appear at a more or less late stage, and that they serve to distinguish monocotyledons from other groups. The plan of the young stele, e. g., Smilacina, bears a close resemblance to that of a dicotyledon, and differs from the older stele of a dicotyledon only in the absence of cambium. The resemblance between the two groups is further shown by the occurrence of medullary strands in several dicotyledonous families, e. g., Nymphaeaceae, and in the older sub-teranean stem of *Ranunculus acris* (6, p. 20); also by the occurrence of amphivasal strands in the mature tissues of such plants as Rheum and Campanula. These considerations lead to the conclusion that the monocotyledons are not an ancient group, but that they have branched off from the dicotyledons, or that both groups have sprung from a parent stock which resembled the modern dicotyledons more closely than it did the monocotyledons.

CONCLUSIONS.

1. The members of the Araceae and Liliaceae have primitively a collateral tubular central cylinder, or ectophloic siphonostele, derived from a protostele, and interrupted by gaps above the points of exit of the foliar traces; through these gaps the external and internal phloeotermas communicate; the intrastelar parenchyma is to be regarded as having the same origin as the cortex, i. e., both cortex and medulla are portions of the fundamental or ground tissue.

2. This primitive condition becomes altered (1) by degeneration of either the internal phloeoterma or both the internal and external phloeotermas; (2) by the assumption of a medullary course by some vascular strands, with which leaf traces are connected; hence the scattered arrangement of bundles is to be regarded as a cenogenetic character.

3. The amphivasal concentric strands are not a palingenetic feature, for they are derived from collateral strands and do not occur in the base of the seedling nor in the leaves and floral axes.

4. Anatomical evidence favors the derivation of monocotyledons from dictoyledonous ancestors.

The subject of this paper was suggested by Dr. E. C. Jeffrey, and the investigation has been carried out under his direction and that of Dr. J. M. Coulter; to both of these gentlemen I wish to tender my best thanks for valuable assistance in the work and in securing material. Thanks are also due Miss E. Sargant, of Reigate, England, for material; to Professor Ikeno, of Tokio, for seeds of Anemarrhena; and to Sir W. T. Thiselton-Dyer, Dr. N. L. Britton, and Dr. W. Trelease for seeds of various species from the botanical gardens under their direction.

THE UNIVERSITY OF CHICAGO.

LITERATURE CITED.

1. ENGLER, A., Beiträge zur Kentniss der Araceæ. V. Bot. Jahrb. **5**:141–188, 287–336. 1884.
2. FARMER, J. B., and HILL, T. G., On the arrangement and structure of the vascular strands in *Angiopteris evecta* and some other Marattiaceae. Annals of Bot. **16**:371–402. *pls. 16–18.* 1902.
3. GWYNNE-VAUGHAN, D. T., On polystely in the genus Primula. Annals of Bot. **11**:317–325. *pl. 14.* 1897.
4. ———, Observations on the anatomy of solenostelic ferns. II. Annals of Bot. **17**:689–742. *pls. 33–35.* 1903.
5. JEFFREY, E. C., The morphology of the central cylinder in vascular plants. Rep. B. A. A. S. **1897**:869.
6. ———, The morphology of the central cylinder in the angiosperms. Trans. Can. Inst. **6**:1–40. *pls. 7–11.* 1900.
7. ———, The structure and development of the stem in the pteridophytes and gymnosperms. Phil. Trans. Royal Soc. London **195**:119–146. *pls. 1–6.* 1902.
8. ———, in Coulter and Chamberlain's Morphology of Angiosperms. 1903.
9. QUEVA, C., Contributions à l'anatomie des Monocotylédonées. I. Trav. et Mém. de l'Univ. de Lille VII. **22**:1–162. *pls. 1–11.* 1899.
10. RIMBACH, A., Ueber die Lebensweise des *Arum maculatum*. Ber. Deutsch. Bot. Gesells. **15**:178–182. *pl. 5.* 1897.
11. SARGANT, E., The origin of the seed leaf in monocotyledons. New Phytol. **1**:107–113. 1902.
12. ———, A theory of the origin of monocotyledons, founded on the structure of their seedlings. Annals of Bot. **17**:1–92. *pls. 1–7.* 1903.

13. SCOTT, R., and SARGANT, E., On the development of *Arum maculatum* from the seed. Annals of Bot. **12**:399–414. *pl. 25.* 1898.

14. SCHOUTE, J. C., Die Stelär-theorie. Jena, 1903.

15. STRASBURGER, E., Histologische Beiträge **3**:—.

16. VANTIEGHEM, PH., Recherches sur la structure des Aroidées. Ann. Sci. Nat. Bot. V. **6**:72–210. *pls. 1–10.* 1866.

17. ——, Sur la polystélie. Ann. Sci. Nat. Bot. VII. **3**:275–322. *pls. 13–15.* 1886.

EXPLANATION OF PLATES XII–XV.

PLATE XII.

FIG. 1. *Acorus Calamus;* section through stele at point of exit of traces of third leaf. × 90.

FIG. 2. Same; part of mature stem: *b*, bundle in process of division into leaf trace and medullary bundle; *g*, gap through which a leaf trace has recently passed; *c*, cortex; *m*, medulla. × 55.

FIG. 3. Same; central cylinder from base of flowering axis. × 35.

FIG. 4. *Symplocarpus foetidus*; section through basal region of tuber. × 15.

FIG. 5. Same at region of the cotyledonary gap (*g*). × 25.

FIG. 6. Same a short distance higher up; cotyledonary gap closed. × 25.

PLATE XIII.

FIG. 7. *Symplocarpus foetidus;* specimen showing early degeneration of internal phloeoterma. × 25.

FIG. 8. Same; region above closure of the cotyledonary gap. × 25.

FIG. 9. *Peltandra virginica;* general view showing cotyledon (*c*) partly separated off; *r, r*, secondary roots. × 7.5.

FIG. 10. Same; basal region of seedling. × 30.

FIG. 11. Same; region of cotyledonary gap; *r*, secondary root. × 25.

FIG. 12. Same; cotyledonary gap closing. × 25.

PLATE XIV.

FIG. 13. *Clintonia borealis;* stele of mature rhizome. × 25.

FIG. 14. Part of section shown in *fig. 13.* × 85.

FIG. 15. *Clintonia umbellata;* stele of mature rhizome. × 20.

FIG. 16. *Maianthemum bifolium;* stele of mature rhizome shortly above region at which it turns upwards. × 25.

FIG. 17. Same at the second node. × 85.

FIG. 18. *Smilacina stellata;* stele in the second internode. × 125.

PLATE XV.

FIG. 19. *Smilacina stellata;* higher region of seedling; *m*, bundle which has left periphery of stele to run for a distance in the medulla; *l*, leaf trace. × 65.

FIG. 20. *Trillium grandiflorum;* young stele: t, median trace of a leaf; t_1, lateral trace of same leaf. $\times$ 30.

FIG. 21. Same; stele slightly higher than the preceding, showing two foliar gaps overlapping. $\times$ 50.

FIG. 22. Same still higher; t_1, median trace; t_2, t_3, lateral traces. $\times$ 50.

FIG. 23. *Lilium canadense;* node of seedling; t, median trace of leaf lying to the right. $\times$ 60.

FIG. 24. *Uvularia grandiflora* immediately above point of exit of the cotyledonary trace. $\times$ 95.

CPSIA information can be obtained
at www.ICGtesting.com
Printed in the USA
BVHW080928211118
533723BV00013B/1124/P